Annual Student Conference: Technical Program

Held at the 2024 AIChE Annual Meeting

San Diego, California, USA
27-31 October 2024

ISBN: 979-8-3313-1691-4

Printed from e-media with permission by:

Curran Associates, Inc.
57 Morehouse Lane
Red Hook, NY 12571

AIChE
120 Wall Street, FL 23
New York, NY 10005-4020

Phone: (800) 242-4363
Fax: (203) 775-5177

www.aiche.org

Additional copies of this publication are available from:

Curran Associates, Inc.
57 Morehouse Lane
Red Hook, NY 12571 USA
Phone: 845-758-0400
Fax: 845-758-2633
Email: curran@proceedings.com
Web: www.proceedings.com

TABLE OF CONTENTS

A Day in the Life of a Process Engineer

Elizabeth Jensen

What does a Process Engineer do at an energy company? What are some of the daily tasks and challenges that they work on? How do they enable lower carbon operations and support the energy transition? What skills (beyond engineering) do they find most valuable? How have they transitioned from an academic to professional environment? Join Process Engineers from Chevron as they describe their roles, share what they have learned, and explain how they have integrated teamwork, communication, technology, and sustainability into their careers.

The Digital Plant of the Future: How Big Data and AI are Affecting the Way We Operate

Adam Gottlieb

As engineers, our ability to perform quality work is directly dependent on the data and the assumptions that we use in our analyses. The simple truth is that wherever you go to work, that data will be fragmented, duplicated, difficult to find, in peoples' heads, in an old dusty file room, missing, or just plain wrong. This workshop will delve into how engineering in manufacturing is generally done now and highlight the potential pitfalls in the current data environment. We will then pivot into a much more interesting discussion about how Celanese has undertaken an ambitious digital transformation to solve many of these problems. Integrating data, using data at scale, creating meaningful data history, and leveraging the latest in artificial intelligence technology are all being used to make our business safer, more sustainable, and streamlined. This workshop will be based around some real examples and will include demonstrations of generative AI tools, 3D modeling, and integrated data platforms.

Beyond the Bachelor's: Mastering Graduate School and Fellowship Applications

Carmen Daoud

Are you an aspiring chemical engineer ready to take the next step in your academic journey? Join us for this informative session where we will dive into the intricacies of applying to graduate school and fellowships. From crafting a compelling personal statement to securing strong letters of recommendation, we will provide you with expert advice and proven strategies to maximize your chances of acceptance. You will also learn about the various funding opportunities available through fellowships and gain valuable insights on how to navigate the application process for them. Whether you're considering a Master's or Ph.D. program, this session will equip you with the tools and knowledge to confidently pursue your academic goals and launch a successful career in chemical engineering.

How To Build a Great Student Chapter

Matthew Liberatore

Growing as a student, engineer, and leader commonly involves being part of larger organizations. Professional engineering student organizations are one example that many engineer students engage with as students as well as through their career. This workshop will lead participants through exercises to help build a great AIChE student chapter this semester and year. By applying an entrepreneurial mindset, participants will identify characteristics desired in their student chapter, explore actions of student leaders, and develop a story of action.

Secure the Future By Applying Process Simulation for Sustainable Process Design

Bruna Giron, Jesse Williamson, Mike Donahue

This workshop will focus on demonstrating how simulation can achieve both environmental and economic benefits in real-world applications.

Chemical Engineering in the Pharmaceutical Industry

Dana Alhasson

This presentation and panelist Q&A will focus on career paths within the pharmaceutical industry for chemical engineers. It will consist of an overview of the industry, multiple career paths (R&D, Manufacturing, Quality, etc.) and will open to questions from the audience to allow students to ask the panelists any questions.

Escape From the Zombie Boss

Alaina Levine

Zombie bosses or colleagues are everywhere – they hide in plain sight and smile at you, but behind the scenes they are working up a storm to bring you down. They may harass you, steal your work, malign your reputation, or worse. How do you escape, survive, and thrive? In this session, Alaina G. Levine will lead us through the various scenarios in which a zombie boss or colleague reveals their true colors so you can be aware of the warning signs. We will then plot your escape – in a professional manner of course, that distances you from the perpetrator while maintaining and even improving your own reputation. It's not always easy to get away from a zombie, but it is important to remember that it is doable and necessary, especially if you want to preserve your mental, physical, and emotional health and advance your own career. You can take ownership, reclaim your power, and get out of that abusive relationship. We will show you how!

From Campus to Corporate: Navigating your Chemical Engineering Career to Success

Ryan Morrison, Lucianna Moos, Eugenia Ndieze

As a chemical engineering student, you are about to embark on an exciting journey into your impending professional career. However, transitioning from college to corporate can be challenging, and it's important to be prepared for what lies ahead. This presentation will provide you with the tips and tricks for excelling in your first year as a chemical engineer after college graduation. We will highlight some technical skills that are essential for success, as well as the importance of networking and building relationships within the industry. Additionally, we will discuss the leadership skills that are necessary for career advancement and how to develop them. By the end of this presentation, you will have a clear understanding of what it takes to thrive in the chemical engineering field and be equipped with the tools to make the most out of your first year after college graduation.

Finances and Your First Job – Financial Planning for ChemE Careers

Carmen Daoud, Shelby Mills

How do I start saving money? What benefits should I enroll in? How do I start planning for retirement? Join the Early Career Community (ECC) for a discussion on the benefits and rewards of financial planning when starting your first job. This session will feature ECC members in a variety of careers (graduate school, industry, government, etc.) and discuss tips for money management and planning for retirement. YPC members will discuss topics such as investing in 401Ks, budgeting, benefits packages, student debt, and planning for retirement. YPC members will also discuss topics that they wish they knew when starting their first jobs.

Become a Chem-E-Car Hero: Best Practices in Chem-E-Car Safety and EDP Development

Jeffrey Heil

In this insightful presentation, attendees will get an in depth view into AIChE's Chem-E-Car program, which has undergone significant safety and operational advancements over the last five years as it has grown from an American event to an International one. The session will commence with a review of the key rule changes that have been implemented to enhance safety and efficiency in the competitions and their relation to industrial standards. Participants will gain valuable knowledge from regional AIChE Safety Coordinators about the most pressing safety concerns in the design and operation of these chemical-powered vehicles.

The talk will explore application of the Globally Harmonized System (GHS) for hazard communication in the Chem-E-Car program. Current and future teams will learn how GHS and other elements of Process Safety Management (PSM) are employed to assess and mitigate chemical risks effectively both during fabrication and at the competition. Emphasizing the need for environmental stewardship and sustainability, the presentation will also cover the protocols for safe waste disposal and the management strategies enforced at both regional and national competitions. Applications to industrial standards are included as appropriate.

Lastly, the audience will be given an inside look at the need for attention to detail required in crafting a comprehensive Engineering Design Package, which is paramount to the success of any Chem-E-Car project. This abstract encapsulates a session that promises to be as educational as it is engaging, offering a thorough overview of the Chem-E-Car program's commitment to safety, innovation, and sustainability.

Learning Objectives:

- Identify key safety and operational rule changes in Chem-E-Car over the past several years and how this relates to industrial practices.
- Review, directly from regional AIChE Safety Coordinators, the most prevalent safety concerns experienced in designing and operating a Chem-E-Car.
- Apply the Globally Harmonized System (GHS) to evaluate and manage chemical risk within the Chem-E-Car competition.
- Develop an understanding of the importance of safe waste disposal, how it is managed at the regional and national levels, and where the guidelines came from (industry perspective).
- Apply an appropriate attention to detail that is required for a well thought out and complete Engineering Design Package (EDP) that includes everything the initial reviewer needs to pass your car.

Living Out Inclusivity in Chemical Engineering and Doing a World of Good in Society

Reginald Rogers Jr.

This will be a moderator-led panel where students and professionals, who identify with DORIC, share personal stories of how they seek inclusion in the Chemical Engineering profession. The moderator with prepare a set of questions for panelists to respond to during the first 20 minutes of the session. Example questions include:

- How have you been able to leverage your challenges to promote a more inclusive environment where you work or go to school?
- What would you like individuals who do not identify with a disability to know about respecting individuals who do have a challenging ability?
- Has there been an instance where you were excluded from an activity because of the perception that you wouldn't be able to be fully engaged in that activity?

The remaining 25 minutes will be reserved for discussion with the audience. The aim will be for audience members to develop ideas for taking back to their home institutions and carry forward into their graduate school or professional careers.

Panel – Composition of panel is targeted to be 2 students and 2 professionals.

Open Discussion – The open discussion portion of the workshop will be structured to allow students to get into small groups to talk about the perspectives they heard from the panelists and come up with ideas for promoting inclusivity at their universities and within the community. Panelists will be encouraged to walk through the audience to hear what ideas students are developing and provide feedback.

The Chemical Engineering and Materials Science Program at Michigan State University

Shiwang Cheng, Robert Ferrier Jr.

The Department of Chemical Engineering and Materials Science at Michigan State University (http://www.chems.msu.edu/) offers MS and PhD degrees in both Chemical Engineering and Materials Science and Engineering. We provide research opportunities in three main thrust areas: Energy and Sustainability, Nanomaterials and Nanotechnology, and Biomedical Engineering & Biotechnology. Within these thrusts lie myriad research activities in diverse fields that include thermoelectric, fuel cells, renewable energy production, hydrogen production, biorenewable polymers, bio-based industrial materials, smart materials, systems biology, proteomics, tissue engineering, and biosensors. Key research laboratories include the Great Lakes Bioenergy Research Center, the Composite Materials and Structures Center, the Interdisciplinary Bioelectronics Training Program, and the Center for Revolutionary Materials for Solid State Energy Conversion. The department offers accelerated online bridging courses and chemical engineering certificates specifically designed to prepare non-chemical engineering students for graduate studies in chemical engineering.

Green Collar Career Opportunities for Chemical Engineers

Chad Able, Matthew Alexander, Kirti Yenkie, Joseph Cramer, David Russell

This workshop is designed to be an interactive discussion with experienced panelists in the environmental field for chemical engineers. The workshop will open with a presentation of opportunities in the environmental field in industry, academia, consulting and government and will conclude with a Q&A session with the panelists to field questions from prospective engineers in this field.

Building a 3-Pillar Outreach Program for Your Student Chapter

Sarah Flannery, Flynn Baliton

Join the University of Florida to learn how to build an outreach program on the pillars of community, K12, and sister chapter relationships. We will walk through our chapter's tried & true strategies for developing meaningful activities, generating excitement & participation within your chapter, and ensuring your events make an impact. Other agenda items will include building long-standing relationships with community organizations and assembling a team & a submission to AIChE's K12 STEM National Outreach competition.

Network Your Society

Alaina Levine

As a member of AIChE, you have a ton of fab benefits and the most underutilized one is the network. As the Global Home of Chemical Engineers, you have the opportunity to find and build win-win alliances with like-minded ChEs across the world, hone your skills, and access hidden jobs and internships, and it all starts with understanding how to leverage your AIChE membership for networking and engagement. In this fun session, you will learn how to network via AIChE to enable your career growth through its vibrant and diverse network. We will start with the fundamental that networking is the most honorable and valuable endeavor in which you can engage, because it is built on a spirit of generosity. We will discuss strategies and tactics for engaging with AIChE and finding new collaborators and building mutually-beneficial partnerships with ChE professionals at in-person meetings and across the globe. You will discover actual wording and questions to use to spark meaningful conversations, and we will also explore networking using digital platforms and how to engage with people you meet at virtual events. You will emerge more confident and comfortable in your networking- now that you know exactly how to do it. Let's network your society!

Imperial College London: Your Future Self Is Waiting for You

Cleo Kontoravdi, Maria Papathanasiou

Interested in taking the next step in your career in one of the world's leading Chemical Engineering Departments? Join us to find out about the postgraduate opportunities offered by the Department of Chemical Engineering at Imperial College London. Situated at the heart of London, ranked the best student city in the world, Imperial College London is a global institution with a multicultural study environment and a vibrant student community.

The event will include information on:

1. the 1-year MSc programme in Advanced Chemical Engineering and the Biotechnology and Process Systems Engineering streams;
2. the PhD programme that covers the broad range of research themes hosted by our Department;
3. the postgraduate admissions process for MSc and PhD studies;
4. the scholarship opportunities offered by our Department.

The event will feature a presentation by the Director of Postgraduate Studies, Professor Cleo Kontoravdi, followed by a Q&A session. It is aimed at current undergraduate and MSc students working towards a degree in Chemical Engineering, Materials Engineering, Chemistry, Biotechnology and other related disciplines who are planning to undertake further technical and research training. It will also be relevant to Engineering and Science professionals wishing to develop their technical and research skills and obtain further qualifications.

AI in the Workplace

Carmen Daoud, Shelby Mills, Prodromos Daoutidis

What is AI? And how does AI affect the work the Chemical Engineers do? As AI is becoming a highly utilized technology in recent years, companies are starting to look for ways to utilize the AI technology in their day-to-day operations to increase efficiency of currently used methods. For example, by utilizing AI, companies can now expand their knowledge base resulting in more refined method development and product testing. Join the Early Career Community (ECC) and CAST (Computing & Systems Technology Division) as we discuss how AI is currently emerging in the Chemical Engineering space.

Self-Care and Resilience

Alaina Levine

One of the silver linings of the pandemic is the fact that it highlighted the need for more self-care, emotional and mental health support, and a greater appreciation for the human experience. After all, we are not AI-driven robots- we think and feel and experience pain, sadness, anxiety and depression, especially in the face of great challenges. In this session, we will discuss strategies for effective self-care. We will delve into ways to navigate any challenging circumstance to protect our physical, mental and emotional health, and you will emerge with a template for a set of guidelines you can develop yourself to ensure you are always kind to yourself.

Path Towards Nanoscale Science and Engineering

Albert Liu

Dive into the fascinating world of nanoscale science and engineering with our expert panel at this year's AIChE Student Conference. This session is designed to introduce students to the exciting opportunities in nanoscale science and engineering, providing a comprehensive overview of the Nanoscale Science and Engineering Forum (NSEF).

During the panel, you will learn about:
1. The foundational concepts and recent advancements in nanoscale science and engineering.
2. Various pathways to get involved in the field, including academic research and industry careers.
3. Insights from experienced professionals and academics who have made significant contributions to NSE.

Don't miss this chance to explore the cutting-edge field of nanoscale science and engineering and discover how you can become part of this innovative community. Whether you're considering a research pathway or looking towards industry applications, this session will provide valuable guidance and inspiration.

Materials Engineering and Sciences in AIChE

Adrianne Rosales

The MESD workshop will feature speakers across all areas of the Materials and Engineering Sciences Division, including polymers, biomaterials, inorganic materials, electronics and photonics, and composites. Representatives from each area will present representative research to give students an overview of research programs in MESD. Speakers will also present advice to students as they pursue and consider different PhD graduate programs or other career options. After the presentations, speakers will be available for a social "meet and greet" with students.

Mixing of Highly Viscous and Non-Newtonian Fluids - What Is Rheology and Why Should We Care?

Clara Gomez, Tanja Tomkovic

Mixing of highly viscous and non-Newtonian fluids - what is rheology and why should we care? This workshop will start with an introduction into complex viscous and non- Newtonian fluids going through the fundamentals of rheology including simple rheology measurement techniques and basic rheological models that describe relevant fluids in our daily lives.

We will then walk the students through the effects that non-Newtonian rheology have on simple fluid flow systems ranging from pipe flow to stirred reactors and multiphase systems of relevance in multiple industrial applications.

Navigating Dynamic Careers in Catalysis and Reaction Engineering

Jesse Bond, Nicholas Brunelli, Christian Jungong, Heather LeClerc, Umar Aslam

Panelists will discuss their careers in the field of Catalysis and Reaction Engineering; how the Catalysis and Reaction Engineering Division has played a role in their careers; and advise on how to navigate a dynamic and fulfilling career in the field. Panelists are selected from academia and industry and can offer insights into these career tracks. The panel is designed to engage the audience and provide opportunities for interactions through extensive Q&A and audience polls.

I Need a Job Right Now

Alaina Levine

Need a job now? You came to the right place! This session will give you a step-by-step strategy to land a job right now, and is applicable to internships and permanent roles in industry, and any other ecosystem. Don't panic!

Beyond Chem-E-Car: Unlocking Hidden Gems of AIChE Competitions for a Winning Career Strategy

Troy Vogel

Need a resume booster? Academic competitions offer far more value than cash prizes and bragging rights. AIChE's carefully curated suite of competitions give you an avenue to practice your presentation skills, critical thinking, applying your ChemE knowledge outside the classroom, and to learn new skills, ultimately making you stand out in a competitive job and grad school market.

You've heard of Chem-E-Car, but what about the Global Undergraduate Student Video Competition or the Undergraduate Student Technical Presentation Competition or the ChemE-Sports™ Competition? In this session, leaders from the AIChE Student Chapters Committee (SCC) and their guests will give you the inside scoop about some of the lesser known AIChE Competitions and what steps you can take to get involved!

Prioritizing Mental Health: A Workshop for Engineering Students

Sarah Wilson

Research shows that engineering students often struggle to find time for their mental health, are hesitant to seek help, and tend to normalize the high stress levels of their training environment. Therefore, this interactive workshop is designed specifically for engineering students to help you prioritize your mental health. We will cover four essential topics:

1. Why Mental Health Matters: Understand the critical importance of mental well-being in your academic and personal life.
2. Advocating for Yourself and Others: Learn how to speak up for your mental health needs and support your peers in doing the same.
3. Recognizing Stress vs. Distress: Gain insights into differentiating between normal stress and signs of mental health issues.
4. Finding Campus Resources: Discover the mental health resources available to you on campus and how to access them.

By attending this workshop, you'll gain valuable skills and knowledge to help you manage stress, seek help when needed, and create a supportive community within your engineering cohort. By learning to prioritize your mental health as an engineering student, you will not only support your current well-being and academic performance but also set the foundation for long-term well-being.

MATLAB and Simulink for Chem-E-Car Competition

Aycan Hacioglu

Every year, chemical engineering students design a shoebox-sized car powered by chemical reactions that must travel a specified distance, as announced on the day of the Chem-E-Car Competition. Are you up to tackling this interesting design challenge with MathWorks tools and resources? We have exciting news for Chem-E-Car Competition Teams! MathWorks is sponsoring the 2024 Chem-E-Car Competition and providing free access to MathWorks tools and curated resources so that you can perfect your car with model-based design. In this webinar, we will highlight how computational tools can help you during your Chem-E-Car design process.

Starting Your First Position in Industry

Carmen Daoud, Shelby Mills

What do employers look for in a recent college graduate? How can you prepare to ace your industry interviews? What does the typical day of an engineer look like in different fields? How can you make the most of your internship experiences to propel you into a full-time job? How can AIChE help you land a job and excel in your first few years out of college? Join the Early Career Community (ECC) in a discussion on preparing to apply for jobs and excel in your first position. This session aims to give the students an idea of what a Chemical Engineer can do in industry – such working in a plant, as a Project Manager, etc. ECC members will also discuss what they wish they knew when starting their first jobs. The session will feature recent-graduate young professional engineers from various companies talking about their experiences and giving advice on making the transition from college to career.

Survive and Advance with Confidence: Navigating the Ph.D. Labyrinth

Reginald Rogers Jr.

A PhD is more than just getting three letters behind your name. It is a journey where you are learning more about yourself as you become an expert in your chosen field of study. However, no two PhD journeys are the same. From Year 1 to Year 5, there will be many crossroads one will encounter before reaching the finish line. Each crossroad may leave you asking: "Is the PhD right for me?" Such a question is common among individuals pursuing the ultimate advanced degree. In this workshop, we will walk through the PhD journey from start to finish. How do you tackle each year of your program? How you negotiate difficult advisors? What do you do when you begin to feel the "PhD Fatigue"? How do you leverage your PhD experience in developing the next phase of your career? The nuts and bolts to tackling a PhD will be the same across the board, but the end result will depend on YOU. This interactive workshop will give you the chance to think about what you need to do in order to be successful and advance to the next level...with confidence!

Learning Outcomes and Take-Aways
By the end of this workshop, participants should be able to:
1. Understand the complexities of navigating a PhD with a different mindset compared to their undergraduate degree program
2. Write out a plan of action for successfully navigating their PhD journey
3. Discuss, freely, issues they are currently facing in their PhD program and how to overcome those issues
4. Walk away confident in their efforts to complete their PhD with a high level of success

Starting Your ChemE Cube Team!

Noel Stanley, Aditi Gali, Harold Ly

Qualifying Requirements
Advertising your team
Having successful meetings
Team leadership

The Path to Becoming a Licensed Professional Engineer

Joseph Cramer, William Parrish, Linda Bergeron

This presentation discusses how becoming a Professional Engineer (PE) can enhance a chemical engineer's career. Students learn about the PE licensure process which includes the Fundamentals of Engineering (FE) Exam and Practice of Engineering (PE) exams, and educational and work experience requirements. The presentation utilizes embedded videos of young PEs and stresses the value of taking the FE exam before, or shortly after, graduation. The workshop leaders are Bill Parrish, Linda Bergeron and Joe Cramer who are members of the National Council of Examiners for Engineering and Surveying's (NCEES) Chemical PE Exam Committee as well as AIChE's Licensing and Professional Development Committee. Together they have a broad range of technical and managerial experience in chemical engineering practice and are Fellows of AIChE.

Process Safety: What Does It Mean for Me?

Noel Stanley

Pillars of Process Safety
Making your MOC the best it can be
How to participate in an incident investigation

Tools, Strategies and Skills to Boost Your Career Development in ChemE

Aura Ramirez

In this workshop we will learn about some tools, strategies and skills that as students we can use to improve our knowledge in Chemical Engineering, get better at interpersonal relationships and explore about what the market needs of us and how we can contribute to it. We will also talk about the importance of climate change and how we can add from our career to a better world.

Creating and Sustaining STEM Ambassadors

Korey Cantrell, Sarah Porter

Learn about the STEM Ambassadors program, designed to engage students in STEM careers through hands-on experiments and mentorship. It provides guidance on how to initiate and sustain the program, including securing funding, building a team, and partnering with schools. This program emphasizes safety, fun, and education in STEM fields.

High School ChemE Car Intro

Reese Rodgers, Ashton Moore, Koen Wertman

Our Goal was to introduce Chemical Engineering to high school students through a safe version of ChemE Car. The way we accomplish this is by replacing the electrochemistry aspect of the car with a battery pack and instead talking about the concepts and examples of electrochemistry and teaching the students about them. Though we create a competition where we introduce the concepts of the much safer clock reaction and have the students learn about and apply the chemistry to compete in a ChemE Car Competition amongst themselves. We want to do this to help market chemical engineering to high school students who may not want to do it because of the lack of physical interaction we have with the materials we use.

Engaging Young Minds: A Guide to STEM Outreach

Victoria Taylor, Catherine Boltz, Katie Evans

This is a comprehensive plan for a STEM Outreach Workshop aimed at engaging and educating young minds about the importance of STEM. The goal of the workshop is to foster a deeper interest in STEM among K-12 students through interactive, hands-on activities and collaborations with various organizations, ensuring a fun and educational experience that highlights the relevance of STEM in everyday life.

Empowering Tomorrow's Innovators: Strategies for a Successful K-12 Program/Event

Riley Coulter, Anna Blobaum

In this workshop we will cover how we started our program, how to plan for experiments, events we have held, how to find funding, how to reach out to classrooms / plan events, and how we equip our chapter volunteers to best assist students in experiment.

Join, Lead, Succeed: Boosting Membership with Technical and Leadership opportunities with AIChE@UCLA

Sergio Curiel, May He

This workshop will explore AIChE at UCLA's Intern Board program and Chemical Engineering Introductory Project (ChIP). While both programs aim to increase first year involvement within the organization, the Intern Board focuses on developing leadership skills whereas ChIP builds technical skills through hands-on engineering. Program structures and achieved results will be discussed.

ChIP: Forming the Foundation of Underclassmen Engineers

Ryan Stearns, Walter Kretzer, Daniel Orrell

Our chapter has pioneered a new program in which underclassmen students design, construct, and operate a chemical engineering process of their choosing. This program incorporates an entry-level approach to complex concepts, solidifying participants understanding of fundamental chemical engineering principles. Furthermore, it presents opportunities for hands-on experiences with process equipment prior to their first internship. We hope for other chapters to adopt this program as well, bettering the quality of education for all participants.

Demystifying Stanford Engineering's Graduate Application Process for First-Generation and/or Low-Income Students

Milenia Rojas Mendoza, Alexandra Ramos Figueroa, Luz Jiménez Ruvalcaba, Wrayzene Willoughby

With a focus on first-generation and/or low-income prospective applicants, this session (1) provides an overview of Chemical Engineering graduate programs at Stanford, (2) demystifies the graduate application process at Stanford Engineering, and (3) offers tips for a competitive application from 3 Stanford Chemical Engineering PhD Candidates.

Chemical Engineering Matters! - Plastics, Climate, and AI

Heather Yuengling, John Cirucci, Jessica Winter, Alexis Voulgaropoulos

Chemical engineers are engaged in a full spectrum of technological arenas with increasing social, economic, and political relevance to our daily lives. AIChE aspires to be a catalyst, applying chemical engineering expertise to meet societal needs and inform science-based policy. As future chemical engineering leaders, we would like to hear your concerns and ideas! Join us for a lightning round discussion on three current hot topics:

• Plastics – discussion leader, Alexis Voulgaropoulos

• Climate – discussion leader, John Cirucci

• Artificial Intelligence – discussion leader, Jessica Winters

This session for students mirrors the Town Hall at the AIChE Annual Meeting.

How to Make the Most of Your Omega Chi Epsilon Chapter

Christi Luks, Troy Vogel, Glenn Lipscomb, Richard Davis

Gather to share ideas for making your chapter the best ever. All officers and chapter members are invited to attend. Schools wishing to start an OXE chapter are also welcome to join us.
Christi Patton Luks, Omega Chi Epsilon President
Troy Vogel, Omega Chi Epsilon Vice-President
G. Glenn Lipscomb, Omega Chi Epsilon Treasurer
Richard A. Davis, Omega Chi Epsilon Executive Secretary

Study for Free - No Tuition Fee - In Germany

Alexander Ditter, Oliver Friedrich

Enrolling at German public universities is unique in that it comes with no tuition fees for undergraduate and graduate studies, not even for international students. This makes studying in Germany very attractive to combine international mobility with a solid university education. The Department of Chemical and Biological Engineering at Friedrich-Alexander-Universität Erlangen-Nürnberg (FAU), the third largest university in Bavaria, offers an international Bachelor's and Master's degree program in "Clean Energy Processes" that especially targets an English-speaking audience. Admission is granted twice per year and studies start in April (summer term) and October (winter term) for graduate students and once in October for undergraduate students. Upon successful completion of a Bachelor's and a Master's degree, FAU also offers competitive PhD programs and fascinating real world application topics to obtain a prestigious German Doctorate in Engineering (Dr.-Ing.). Besides general information on tuition free studies in Germany, we will focus on the specific study course majors and research priorities at FAU and provide background about similarities and differences between studying in Germany vs. the United States.

The World of Particle Technology

Ben Freireich

Billions of pounds of bulk solids are processed and handled every year by the US process industries, yet most chemical engineers are ill-equipped to deal with the complexities of engineering science of solids processing. Hence, plants and products suffer with lost production, inability to achieve design production rates, off-grade or off-specification products. During this session, we will take a look at the fun and exciting world of solids processing. Specifically, we will look at some of the more common particle-based technologies examining both the important role they play in society today along with the associated technical challenges.

AUTHOR INDEX